# a FORTUNE in SCRAP

## *SECRETS OF THE SCRAP METAL INDUSTRY*

by

Ken Burtwell
Industry Veteran

Prior edition © Copyright 1983
DKS Publishing by Maudland Press

Copyright © 2012 Ken Burtwell

All rights reserved.

ISBN-10:1480282839

# DEDICATION

This book is dedicated to my wife, Dode.

# CONTENTS

|  |  |  |
|---|---|---|
|  | Acknowledgments | i |
| 1 | Introduction and Disclaimer | Pg 1 |
| 2 | Steel | Pg 4 |
| 3 | The Scrap Dealer or Collector | Pg 8 |
| 4 | Flame-Cutting Steel Scrap | Pg 11 |
| 5 | Better Grade – More Value | Pg 15 |
| 6 | Metal Scrap from Demolition | Pg 16 |
| 7 | The Thermic Lance (aka Thermal Lance) | Pg 27 |
| 8 | Railroads and Rail Vehicles | Pg 29 |
| 9 | Reclamation Equipment | Pg 31 |
| 10 | Sheet Steel Gauges and Weights | Pg 39 |
| 11 | More Valuable Scrap Metals | Pg 45 |
| 12 | Common Metals: Lead and Zinc | Pg 69 |
| 13 | Less Common Metals and Where to Find Them | Pg 72 |
| 14 | Identification of Metals | Pg 75 |
| 15 | Magnetic Testing | Pg 78 |
| 16 | Government Scrap Sales | Pg 79 |
| 17 | Precious Metals | Pg 81 |
| 18 | Silver Coins | Pg 84 |
| 19 | Testing Various Metals | Pg 87 |
| 20 | Scrap Dealers | Pg 89 |

ACKNOWLEDGMENTS

Thanks go to Susan and Alan Gast
of Beesville Books
(www.beesvillebooks.com)
for their proofreading and layout work.

# 1
# INTRODUCTION

By Ken Burtwell

## *Decision Made...*
## *TAKE NO MORE ORDERS*
## *from others*

I was stationed In Palestine in 1946 there was no ISRAEL, and Palestine was a British Protectorate. In the post war armed forces environment it was...

*"Yes Sir", "No Sir"...*

*"If it moves Salute It ... if it DOESN'T move - Paint it White"*

I'd spent two years as an Airborne Motor Cycle Dispatch Rider attached to the 6th Airborne Division. In training I had parachuted 13 times, each time with a different objective, mainly to create what was known as "beach-head" which was a "Command Post" for Operations.

The only time this happened was when we were "dropped" near Tel Aviv when my job was NOT to carry "dispatches" by an Airborne Motorcycle but to jump out of a plane with a battery (for the radio operators) strapped to my right leg - something I had never been trained to do, but that didn't matter as I got ten minutes training on the flight to Tel Aviv.

There is a special way, it turned out, to parachute *with a heavy lead battery strapped to your leg*. This special way was YOU WERE FIRST IN LINE so that on reaching the DZ (drop-zone) YOU WERE "HELPED" out of the door with the other 19 Paratroopers standing in line behind YOU.

We had only "seconds" when the GREEN Light came on for all 20 of us to EXIT the plane...

I didn't need their help - I managed to swing my battery-laden right leg out of the door first which pulled me out of the door. Then, on the way down, I was to pull a pin that let the battery hang 20 feet below me and of course hit the ground up to 20 feet away from me.

The whole operation was a NON-EVENT. We took command of the main highway to the entry to Tel Aviv and nothing happened, after a few days of inactivity we went back to our base camp in Haifa.

When I returned from separating Jews and Arabs in Palestine to be "demobbed" (demobilized) names used at that time for a returning Service man, I decided NEVER to take ORDERS from anyone ever again.

Car Dealing was good business because of the "after the war shortage" but that didn't last so the next best was SCRAP DEALING. It was the same as before, SCRAP was in DEMAND... There was also a "Hungry Market" for raw materials.

All you had to do is

**FIND** it... **COLLECT** IT

And **DELIVER** IT...

*NOTHING HAS CHANGED...*
*FIND a HUNGRY MARKET AND FEED IT!*

## DISCLAIMER

The information contained within this book is no guarantee that the information provided is correct, complete, and/or up-to-date. You use the information at your own risk.

The materials contained in this book are provided for general information purposes only and do not constitute legal or other professional advice on any subject matter.

A Fortune In Scrap does not accept any responsibility for any loss which may arise from reliance on information contained in this book.

The contents of this book are protected by copyright under international conventions and, apart from your purchase, the reproduction, permanent storage, or retransmission of the contents of this book is prohibited without the prior written consent of the A Fortune in Scrap's Copyright holder.

The typewritten links within this book lead to websites which are operated and maintained by third parties. A Fortune in Scrap includes these links solely as a convenience to you, and the presence of such a link does not imply a responsibility for the linked site or an endorsement of the linked site, its operator, or its contents (exceptions may apply).

This book and its contents are provided "AS IS" without warranty of any kind, either express or implied, including, but not limited to, the implied warranties of merchantability, fitness for a particular purpose, or non-infringement.

Reproduction, distribution, republication, and/or retransmission of material contained within this book are prohibited unless the prior written permission of A Fortune in Scrap's Copyright holder has been obtained.

# 2
## STEEL

Steel is produced by various processes and almost always has a percentage of scrap in its content.

It is melted and poured from furnaces into ingot molds. These ingots are solid lumps of steel weighing sometimes a few tons each, and while still red-hot, are put through a series of rollers which reduce the section and increase the length.

Flat rollers will produce plate and sheets and specially shaped rollers will produce constructional steel sections such as beams, channels, angles, and tees.

The manufacturer, engineer, or constructor then turns the finished steel into a useful item of machinery, plant, or building.

After a useful life, the scrap dealer takes over and reverses the process, breaking down to furnace size and separating or segregating the various metals and grades so that the resulting melt will be suitable for making similar items again.

This is an example of RECYCLING, and is a necessary process both from an economical and environmental point of view. There is also a reduced demand on natural resources.

## MAIN TYPES OF FURNACES

- Basic Open Hearth
- Basic Oxygen
- Blast Furnaces
- Electric Furnaces

Some are fed by inserting charging boxes packed with scrap. These are emptied inside and removed and refilled rather like large spoons. The scrap size is limited to the size of the charging box, five feet by two feet, and should be cut to ensure compactness when loaded into the box. A list of different specifications and sizes appears later. One Pittsburgh steelmaker uses 100% scrap, while others use as low as 26%.

The scrap dealer must prepare his scrap, that is, segregate and cut to furnace size to command the best price, or he accepts less and sells it to a dealer who processes the scrap to make it ready for the furnace.

It is almost always the policy to cut it to size on site if large quantities have to be transported to get capacity loads. The scrap is cut (actually it is melted) by an oxygen/propane torch. Long lengths of scrap which can be loaded and transported economically are best sold as shearing scrap to a scrap processor who loads it into his machine's charging box; this moves forward at a pre-set distance depending on the grade of scrap he is making. The blade comes down and actually cuts it to the correct length.

A processor with this equipment usually has a press to bundle light steel scrap for ease of transport and so that it can be charged into the furnace. Movies often show whole cars placed into these huge presses or crushers, but in actual fact, a whole car crushed into a bale would be virtually unsalable because of all the non-steel items present.

## THE FOLLOWING ITEMS HAVE TO BE REMOVED BEFORE AN AUTOMOBILE BODY IS BUNDLED:

- All glass in the windows and lights
- Engine
- Wheels
- Axles
- Entire dashboard

- All wiring including the battery ground
- All rubber hoses and seals
- Die-castings such as grilles, door handles
- All seats, mats, carpets, headlining

Windows must not be broken out as broken glass gets into inaccessible places and cannot be removed. In fact, everything should be removed before bundling that is not steel.

A similar process is performed with refrigerators, stoves, etc. This is bundled separately because of its enameled coating and is of a lower grade.

# 3
## THE SCRAP DEALER OR COLLECTOR

The SCRAP DEALER'S JOB is a never ending search for anything made of metal because he knows that every piece of metal he can lay his hands on is as good as money in the bank. It takes no hard selling as buyers are everywhere, competing for whatever supplies are offered. Location and transport are to be considered, as a selection of prices in different parts of the country show below.

Prices shown reflect prices in 1980… for current prices please visit any of these sites:

        http://www.scrapmonster.com
        (a comprehensive Global site)

        http://www.metal-pages.com/
        (a comprehensive Global site)

        http://www.recycleinme.com/
        (a comprehensive Global site)

        http://www.lme.com/home.asp
        (the London Metal Exchange)

        http://www.scrappricebulletin.com/MarketReportsRegion.aspx?marketcode=LS

The last link takes you to a bulletin of scrap prices, from these cities: Pittsburgh, Chicago, Philadelphia, St. Louis, Detroit, Cleveland, Cincinnati, Youngstown, Buffalo, Birmingham, South Carolina, New York, Boston, Los Angeles, San Francisco, Seattle-Portland,

Hamilton (Ontario), and Houston.

Please note that just about all sites charge a monthly subscription fee for current-day prices.

PRICES FOR GRADE NO. 1 STEEL SCRAP, PER TON (2000 lbs) in 1980

|  |  |
|---|---|
| Philadelphia | $54 |
| Cleveland | $51 |
| Pittsburgh | $49 |
| San Francisco | $46 |
| Birmingham | $45 |
| Houston | $45 |
| St. Louis | $44 |
| Los Angeles | $43 |
| Boston | $37 |
| Detroit | $34 |
| Buffalo | $30 |
| Cincinnati | $30 |

I provide these *dated prices* simply to show you the price differences per ton which can be attributed to the *proximity* of consuming steelworks.

TRUCKERS make money hauling from areas of low demand to consuming steelworks as the variation shows. A scrap dealer with his own trucks can sell in the best markets guaranteeing delivery and profits.

No. 1 grade steel scrap is less than five feet long and is generally at least 1/4 inch thick.

No. 2 grade steel is the same size but less than 1/4 inch thick, minimum of 1/8 inch thick.

The newcomer to the scrap metal business will find a magnet the most useful tool he has. A pocket magnet will tell him that if the metal does not attract, then it is not iron or steel and is more valuable. It is useful in checking various trailers and bus bodies. They could be aluminum or stainless steel (some stainless is magnetic, see further in book). Electro-magnets are used on a crane hook for moving and loading iron and steel scrap.

Electro-Magnet

# 4
## FLAME-CUTTING STEEL SCRAP

Methods of manufacturing articles from steel vary for a number of reasons:

1. When the article was made

2. Where it was made, and

3. Its finished use

Old bridges, ships, and boilers had their sections riveted together and in the old days, if they had to be dismantled, the rivet heads were sheared off and the rivets punched out. With the manufacture of gases for commercial use, more of these parts were welded together instead of being riveted.

Welding means the joining of two pieces of metal by melting so that the edges of the two pieces run together and on cooling are joined as one.

FLAME CUTTING reverses this process; the metal is brought to melting point only in the place where the cut is to be made. This is done with a torch similar to the one shown on the next page.

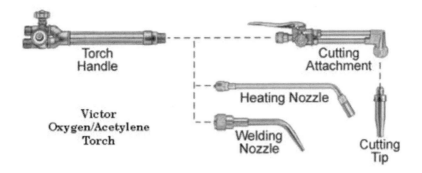

Graphic: McMaster-Carr

This type of torch will cut steel up to ten inches thick depending on the size of the nozzle used.

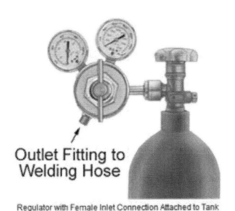

Graphic: McMaster-Carr

The gas producer compresses the gas into steel cylinders at a high pressure so that when we come to use it, we must use a regulator, which is screwed onto the cylinder.

The two pipes from our torch connect to the two regulators on the cylinders, which reduce the pressure for our use.

Two gases are piped into the torch, and there are three control valves. The valve controlling the propane gas is turned on first and is ignited. This is not a hot enough flame so the oxygen supply valve is then turned on slowly. The flame will turn from a yellowy color, grow fiercer, and when a small blue center appears in the flame you are ready to commence flame cutting.

The hottest part of the flame is just above the blue center, so the torch is held so that the tip of the blue flame just touches the steel. In a few seconds the steel begins to melt, but this is not sufficient to cut. The large lever is pressed and a blast of oxygen clears away the molten metal. Care must be taken to prevent fires. GLOVES and GOGGLES must be worn.

This procedure works a little better with wrought iron, but with cast iron more heat is required. Acetylene gas is used instead of propane; (and many of you may have already heard the term "Oxy-Acetylene Torch" used by others). This gives a greater heat initially which helps when the molten metal is blasted away with the oxygen control lever.

Cast iron can also be identified by applying a torch to the surface. Small sparks crackle and rise in the air, this does not happen with mild steel. Mild steel is common steel, used in most fabrications.

Copper, brass, aluminum, and other non-ferrous metals can be cleaned of steel screws, bolts, and other attachments with the use of a torch, but they cannot be cut in the same way that steel can. The metal just melts and must be allowed to fall away.

~*~

*Not ALL scrap GOES for scrap!* The handsome locomotive on the next page is named Lindsay, and was purchased (in the mid '70's) for only $2,000 and sold to a museum for $8,000 in its untouched state – check out their fantastic restoration!

Lindsay "before"

This is what Lindsay looks like today!

# 5
# BETTER GRADE – MORE VALUE

As we get away from domestic scrap, the values increase. Trucks and vans have heavier chassis and engines. Buses, vans and trailers may also have aluminum bodies and sub-chassis.

Farm machinery cuts up easily into wrought iron and steel scrap, at the same time, the cast iron can be separated.

A scrap dealer could visit garages, workshops, repair shops, and farms. More often than not, the owners would say if they were phoned, that they had no scrap. A good collector would be able to show the owner what he is seeking; the owner will have overlooked that pile of JUNK out back.

Machinery and machine parts soon disappear from sight when the grass starts to grow and many a collector has got a good week's pay from what appeared to be a green field or a corner overgrown with weeds. This is more noticeable around farms and repair shops.

# 6
## METAL SCRAP FROM DEMOLITION

Most commercial buildings contain structural steel, some more than others. Warehouses need extra steel in the floors to support the loads they carry. Machine shops also need extra steel to support the weight of the machinery, even the single story workshop or warehouse needs steel to hold up the roof.

Sheets of lead, copper, zinc, or aluminum are used as flashing to keep out the rain. Some heavy industrial shops have iron or steel floors, steel-supported overhead cranes and steel roof trusses. A crane such as this guarantees the existence of copper cable, sometimes without insulation, switch gear, and electric motors - all containing copper.

There will also be cables and transformers and we cover these later in the non-ferrous section.

There are also tables which tell us the weight per foot of steel beams etc. A person can easily estimate the weights of steel in any structure which will enable him to buy the building for demolition and profit. All he has to do is to measure the lengths of the different beams, angles, tees, channels, etc. and measure the section.

Some of these sections will be re-usable and will command a cash price higher than scrap. Sometimes buildings are demolished in such a way that they are damaged or bent and will be cut for scrap on site. The straight re-usable pieces can be salvaged where it is economical. Cranes could support them when they are being cut out.

Re-usable pipes and plates are also salvaged. Some steam boilers have good plate sides that can be cut out and sold for re-rolling. The same would apply to steel ships and storage tanks.

The riveted or welded joints are cut to five feet and make good No. 1 grade scrap for the steel mills.

Chemical plants, heavy machine shops, and shipyards are all good sources of heavy iron and steel scrap.

The really interesting places, from a scrap dealer's point of view, are electric generating plants, which, as well as containing the alternators, contain boilers, condensers, turbines, transformers, circuit breakers, pumps, cables, bus-bars, and motors.

Oil, gas and chemical installations not only produce good iron and steel scrap, but can include lead, copper, brass, and stainless steel.

Part of a Sulfuric Acid plant being demolished

Remains of a Heavy Machine Shop

Also recoverable from chemical installations are most varieties of piping, from stainless through copper, lead, brass, and nickel.

Heat exchangers and condensers all contain valuable tubes and tube plates. Rectifiers can contain copper and mercury; catalytic emission controls can contain platinum.

While we are in the realm of precious metals, photographic materials contain recoverable amounts of silver. Silver can also be found in electrical contacts, while old computer parts can contain gold - more about those later.

## DEMOLISHING AN ENGINEERING MACHINE SHOP

The demolition of an engineering machine shop if the contents are included, can supply lathes, boring, milling, and dozens of other types of metal working machines. All are constructed of heavy cast-iron, with electric motors, steel shafts and bolts, and sometimes bronze bearings.

Electric welding machines have steel-cored copper coils; spot welders have heavy copper arms and contacts.

## VALUABLE SCRAP FROM DEMOLITION

The previous photos of the engineering shop demolition and the photo on this page give an idea of the scope for the recovery of steel scrap from industrial properties. (Prices mentioned circa 1980).

Over 3500 tons were recovered from this redundant iron foundry, worth over $315,000.

Before the roof was dropped, 110 tons of lead was stripped from the gutters and glazing bars, worth $55,000.

Expenses and costs were around $180,000.

**This leaves a profit of around $190,000.**

This huge steel-framed roof was dropped to the ground by burning though each truss, one operator on each side, both cutting the same truss at the same time. This progressively but gradually brought down the roof slowly and safely.

650 tons of steel were recovered **worth over $59,000.**

This dockside crane shown in the photo below was one of dozens that was pulled over and cut up to make the best steel-making scrap.

The photo below shows the toppled crane on the ground and the two burner-operators (look closely at the circled area on the lower left!) who quickly reduced it to furnace size. The purchase price was about $2,000 for each crane. Wages and expenses totaled approx. $5,000 per crane.

170 tons of steel, 15 tons of cast iron and 1100 pounds of copper and brass were recovered **worth over $17,000.**

## BRIDGES

Pictured side-by-side above, are two redundant railroad bridges; one over a side-road, and one over a canal.

Between them they produced 340 tons of Grade #1 steel scrap **worth over $30,000.**

Intermediate beams being cut out with a torch

## LOCOMOTIVES

An old steam locomotive is coming up next (apologies for the lack of photo clarity, they are all over 35 years old). The boiler has been opened up (see circled area of the photo) showing the steel tubes and the copper fire-box from the old locomotive.

98 tons of steel and iron, and 3 ½ tons of copper and brass were recovered

Heat treatment plants in tool shops have valuable heat-resisting equipment containing nickel. Plating plants have anodes and sludges which contain valuable metals. There is also copper in the electrical side of the operation.

Breweries and bottling plants, especially the old breweries, contain good weights of copper, brass, bronze, lead, and stainless steel. Some of the huge vats and brewing vessels are made of copper - worth thousands of dollars - just to be broken up and sold over the scale. You will now have noticed that everything is dollars cash per pound or ton.

There is an instant demand and buyers can be found in every city. The only worry or consideration is: Will the better price I am offered in another city out-weigh the cost of transporting it there?

## SHIP YARDS

Ship breaking can be very profitable, apart from the obvious steel shell, framework, and decks; almost all fittings have to be bronze, as well as pipes and tubes of copper to withstand the wet conditions. Most of the world's tonnage of redundant ships ended up in Japan, Taiwan, Pakistan, Spain, and other countries, but the smaller ones as well as government ships, can still be purchased locally.

The U.S. Government sells thousands of tons of ships and other scrap every month. (Agency addresses further on in the book.)

Another form of transport that becomes available for recycling is the aircraft, but we will deal with that later because very little steel or ferrous is involved.

Scrap propeller and part of shaft suspended from crane in shipyard

This huge propeller and part of the shaft has been flame cut from the ship floating in the harbor. As the superstructure is cut away and lifted ashore, the hull rises in the water so that the propeller can be removed easily.

As more of the hull is cut away, it continues to rise until a saucer-like section of the hull is left, barely able to keep the heavy keel afloat. It is then towed to a place where it can be beached, (preferably on a high tide); there it can be reduced to furnace size.

The shaft will be steel but the propeller could be cast iron or an alloy of copper and zinc with a small percent of manganese - this is known as "manganese bronze".

The operator in the above photo is using a flame cutting torch. He is working on the top section of the bow of the ship in the background. Another torchman will have cut the piece from the inside. A crane then lifts the large sections ashore where they are reduced to furnace size.

Scrap steel from sources such as this is eagerly sought after. The steel mill knows the quality of the melted scrap so that it can be rolled into plates or beams for re-use, probably to make another ship.

# 7
## THE THERMIC LANCE
## (AKA THERMAL LANCE)

The Thermic Lance (also known as the Thermal Lance) is used to penetrate the toughest and thickest of materials. It is used to reduce to furnace size large steel shafts, rollers, rotors, etc. Some castings which cannot be broken are also reduced by this awesome process.

A small bore steel tube of electrical conduit size, about 10 feet long, is filled with similar lengths of steel wire. It is packed tight so that the wire does not fall out if the tube is shaken.

Two or three cylinders of oxygen are connected with copper pipes to a regulator similar to the ones used for burning smaller scrap. The regulator has an adjustable hand screw and it is able to take the pressure from all the cylinders at the same time. The outlet from this regulator is connected to a heavy rubber hose which is also connected to one end of the wire-packed tube or LANCE.

Once this is done, the oxygen is turned on at the cylinders and the THERMIC LANCING process begins. A hole is made with the blast of oxygen and the end of the lance is used to penetrate deeper and to lengthen the cut as required.

The floor underneath the cutting area should be prepared before work is started to prevent fires spreading as the molten metal drops to the floor. Steel plates and plenty of water should be made available.

The end of the lance will melt as the process continues, so that when the operator is positioned too close to the work, a new lance must be ready to be fastened to the rubber pipe and the short one discarded. Regular hose clips are used to connect the two together.

The THERMIC LANCE will blast away through the thickest of steel armored walls or doors, even composite sections where non-metallic substances are positioned between metal plates. On some construction sites the lance has been used to blast holes through concrete floors and walls for the positioning of air conditioning ducts, etc.

If used in a confined space, breathing equipment is needed because of the smoke and fumes.

The process can be very useful as it does not need complicated or expensive equipment. LANCES can be made up from scrap. An oxygen manifold can be purchased or assembled to use with the cylinders (which are rented), along with a regulator, heavy duty rubber hose, and a shield to protect the operator, are all that is required.

# 8
## RAILROADS AND RAIL VEHICLES

One of the biggest producers of heavy scrap is the railroads. The rails are one of the best grades because they are always clean from wear, uncontaminated, and can be easily melted and converted to new rails.

*Rails wear very quickly on curves, around points and crossings, and on downhill stretches where there is much braking.*

There is a constant renewal in progress on well used tracks; worn lengths are burned out and replaced with new. Sometimes the new lengths are welded to the old, and occasionally bolted together with plates.

Locomotives, trucks, cars, and box cars are continually being replaced - the old ones headed for the breakers yard.

There are over 40 grades of railroad scrap. Steel tires are burned off the wheels, and axles are cut on the inside of the wheels.

Brake shoes are usually cast iron but some are contaminated with a composition and can only go to certain foundries.

Couplings and knuckles and malleable castings also find a different address. Malleable castings look like cast iron but are softer and do not break with a knock or hammer blow.

Specification #24 would cover the rest of the railroad car or locomotive. It is: No. 1 Railroad Melting Steel; clean wrought iron or steel scrap ¼ inch and over in thickness, not over five feet long and 18 inches wide, may include pipe ends and thinner material if no bigger than 15 inches by 15 inches. All pieces are cut so as to lie flat in the charging box.

Springs are another grade which can be sold separately.

Rails not too worn can be sold as RE-ROLLING RAILS; these are cut into standard lengths and heated to a red-heat. Then they are rolled into smaller rails or other sections such as tees, angles, and channels.

Worn rails are almost always burned or sheared down to two feet and command a premium price at the mills and foundries.

A rail is nicked with a torch (it is not necessary to burn right though) which means that a groove is cut or burned in the top of the rail and the torchman moves on down the track.

Later another man follows with a sledge hammer, and after the rail has cooled, he gives it a blow, breaking it below the groove.

The best time is a cold winter's morning, apart from it being easier to use a hammer, the cold weather makes the rail more brittle.

The rail around points and crossings sometimes are made of steel to which manganese has been added, this toughens it for the heavy wear that takes place on these sections of the track.

Certain metals that are not iron or steel (non-ferrous), such as copper, brass, bronze, and aluminum, are encountered when wrecking locomotives and cars and are much more valuable. We will cover these later. In the meantime, they are best left under lock and key, as they are as good as ready cash.

# 9
# RECLAMATION EQUIPMENT

Claw crane moving scrap at a scrap yard

Alligator shears left, and pre-compression shears, right

Model of an hydraulic scrap shearing machine

This model of an hydraulic scrap shearing machine, shown above, has side wings for pre-crushing light or bulky scrap before it is sheared. The operator stands at the left of the machine near the top of the stairs. A crane with an electro-magnet loads the charging box, and also clears and loads the furnace size scrap which falls away to the right.

A photo of the real machine showing the right side in action

The photo above shows the sheared scrap after it falls away from the mouth of the shearing machine.

## ELECTRO-MAGNETS

Electro-magnets have a copper or aluminum coil wound around a central core. The electricity is supplied by a generator on the crane, and can be switched on and off by the operator. The outer shell is usually cast from steel.

Electromagnet picking up steel scrap

## TIRE REMOVER MACHINE
## AND HYDRAULIC GRABS

This trailer-mounted machine above makes quick work of separating scrap tires from auto wheels. Mountains of these are found at any wrecking yard.

HYDRAULIC GRAB handling a mixed load of scrap

Most of the scrap in the previous photo could be cut to furnace size on this alligator shear, shown below.

The alligator shears are usually powered by diesel engines or electric motors.

The reversible blades each have four cutting edges.

Alligator Shears

Baling machine

This bushelling or baling machine, above, is working in a large scrap processing yard.

The bales of scrap can be seen leaving the machine on a roller conveyer at the bottom left of the machine. An electro-magnet is loading the machine.

A DIESEL HYDRAULIC POWERED crane, above, equipped with an electro-magnet working in a scrap yard.

*Below, a very special demolition worker – my son, Alan!*

# 10
## SHEET STEEL GAUGES AND WEIGHTS

### Sheet Steel Gauges and Weights

| Gauge | Approximate Thickness Fractional inches | Decimal inches | Millimeters | Weight per sq ft oz avdp | Weight per sq ft lb avdp | Weight per sq meter (kg) |
|---|---|---|---|---|---|---|
| 000 000 0 | 1/2 | 0.5 | 12.7 | 320 | 20.00 | 97.65 |
| 000 000 | 15/32 | 0.468 75 | 11.9063 | 300 | 18.75 | 91.55 |
| 000 00 | 7/16 | 0.4375 | 11.1125 | 280 | 17.50 | 85.44 |
| 0000 | 13/32 | 0.406 25 | 10.3188 | 260 | 16.25 | 79.33 |
| 000 | 3/8 | 0.375 | 9.525 | 240 | 15.00 | 73.24 |
| 00 | 11/32 | 0.343 75 | 8.7313 | 220 | 13.75 | 67.13 |
| 0 | 5/16 | 0.3125 | 7.9375 | 200 | 12.50 | 61.03 |
| 1 | 9/32 | 0.281 25 | 7.1438 | 180 | 11.25 | 54.93 |
| 2 | 17/64 | 0.265 625 | 6.7469 | 170 | 10.625 | 51.88 |
| 3 | 1/4 | 0.25 | 6.35 | 160 | 10.00 | 48.82 |
| 4 | 15/64 | 0.234 375 | 5.9531 | 150 | 9.375 | 45.77 |
| 5 | 7/32 | 0.218 75 | 5.5563 | 140 | 8.75 | 42.72 |
| 6 | 13/64 | 0.203 125 | 5.1594 | 130 | 8.125 | 39.67 |
| 7 | 3/16 | 0.1875 | 4.7625 | 120 | 7.50 | 36.62 |
| 8 | 11/64 | 0.171 875 | 4.3656 | 110 | 6.875 | 33.57 |

## Weights of Steel Plate, I-Beams, Wide Flange Beams and Channels

### I Beams

| Beam Height (inches) | by | Flange Width (inches) | Weight (Lbs. per foot) |
|---|---|---|---|
| 3 | " | 2.375 | 5.70 |
| 4 | " | 2.625 | 7.70 |
| 5 | " | 3.00 | 10.00 |
| 6 | " | 3.375 | 12.25 |
| 7 | " | 3.625 | 15.30 |
| 8 | " | 4.00 | 18.40 |
| 10 | " | 4.625 | 25.40 |
| 12 | " | 5.00 | 31.80 |
| 12 | " | 5.25 | 40.80 |
| 15 | " | 5.50 | 42.90 |
| 18 | " | 6.00 | 54.70 |

## Wide Flange Beams

| Beam Height (inches) | by | Flange Width (inches) | Web Thickness (decimal inches) | Weight (Lbs. per foot) |
|---|---|---|---|---|
| 4 | " | 4.00 | 0.280 | 13.00 |
| 5 | " | 5.00 | 0.240 | 16.00 |
| 6 | " | 4.00 | 0.215 | 9.00 |
| 6 | " | 4.00 | 0.230 | 12.00 |
| 6 | " | 6.00 | 0.230 | 15.00 |
| 6 | " | 4.00 | 0.260 | 16.00 |
| 6 | " | 6.00 | 0.260 | 20.00 |
| 6 | " | 6.00 | 0.320 | 25.00 |
| 8 | " | 4.00 | 0.170 | 10.00 |
| 8 | " | 4.00 | 0.230 | 13.00 |
| 8 | " | 4.00 | 0.245 | 15.00 |
| 8 | " | 5.25 | 0.230 | 18.00 |
| 8 | " | 5.25 | 0.250 | 21.00 |
| 8 | " | 6.50 | 0.245 | 24.00 |
| 8 | " | 6.50 | 0.285 | 28.00 |
| 8 | " | 8.00 | 0.285 | 31.00 |
| 8 | " | 8.00 | 0.400 | 48.00 |
| 10 | " | 4.00 | 0.230 | 15.00 |
| 10 | " | 5.75 | 0.300 | 30.00 |

## Wide Flange Beams (continued)

| Beam Height (inches) | by | Flange Width (inches) | Web Thickness (decimal inches) | Weight (Lbs. per foot) |
|---|---|---|---|---|
| 10 | " | 8.00 | 0.315 | 39.00 |
| 10 | " | 10.00 | 0.340 | 49.00 |
| 10 | " | 10.00 | 0.420 | 60.00 |
| 10 | " | 10.50 | 0.470 | 68.00 |
| 12 | " | 4.00 | 0.200 | 14.00 |
| 12 | " | 4.00 | 0.260 | 22.00 |
| 12 | " | 6.50 | 0.260 | 30.00 |
| 12 | " | 8.00 | 0.295 | 40.00 |
| 12 | " | 8.00 | 0.370 | 50.00 |
| 12 | " | 10.00 | 0.345 | 53.00 |
| 12 | " | 12.00 | 0.390 | 65.00 |
| 12 | " | 12.00 | 0.470 | 79.00 |
| 14 | " | 5.00 | 0.255 | 26.00 |
| 14 | " | 6.75 | 0.270 | 30.00 |
| 14 | " | 6.75 | 0.310 | 38.00 |
| 14 | " | 8.00 | 0.305 | 43.00 |
| 14 | " | 8.00 | 0.370 | 53.00 |
| 14 | " | 10.00 | 0.375 | 61.00 |
| 14 | " | 10.00 | 0.450 | 74.00 |
| 16 | " | 5.50 | 0.250 | 26.00 |
| 16 | " | 7.00 | 0.380 | 50.00 |
| 16 | " | 10.25 | 0.455 | 77.00 |
| 18 | " | 6.00 | 0.315 | 40.00 |
| 18 | " | 7.50 | 0.450 | 65.00 |
| 21 | " | 6.50 | 0.380 | 50.00 |
| 21 | " | 8.25 | 0.430 | 68.00 |
| 21 | " | 9.00 | 0.499 | 82.00 |
| 24 | " | 7.00 | 0.430 | 62.00 |
| 24 | " | 9.00 | 0.470 | 84.00 |
| 24 | " | 12.00 | 0.510 | 117.00 |
| 27 | " | 10.00 | 0.490 | 94.00 |
| 33 | " | 11.50 | 0.550 | 118.00 |

## Weights of Steel Plate

| Thickness (inches) | Pounds per Sq. foot |
|---|---|
| 1.000 | 40.80 |
| 0.500 | 20.40 |
| 0.375 | 15.30 |
| 0.250 | 10.20 |
| 0.125 | 5.10 |

## TWO-PART FORMULA FOR CALCULATING THE WEIGHT OF A CIRCULAR TANK

To find the weight of a **circular tank, <u>two</u> formulas** are used:

**1)** Pi multiplied by the radius squared ($\pi \times R^2$) = area of top or bottom

AND

**2)** Pi multiplied by the diameter by the height ($\pi \times D \times H$) = area of the side walls

The symbols above are represented by the following:

$R$ = radius, or half the diameter
$D$ = diameter, or distance across the circle
$X2$ = number squared, or the number multiplied by itself (example: 5 x 5 = 25)
Pi ($\pi$) = 3.14159

Example:

> Height of tank - 10 ft.
> Diameter of tank - 5 ft.
> Thickness of tank - **0.375** inches
> Tank has <u>top</u> and <u>bottom</u>.

**Formula #1** - $\pi \times R^2$ = 3.14159 x (2.5 x 2.5)
= 19.635 square feet (area of top of tank)

Area of top AND bottom = 19.635 <u>x 2</u>
= 39.27 square feet

39.27 x 15.3 pounds (15.3 taken from chart; weight of 0.375 steel plate)
= **600.83 pounds** (weight of top and bottom of tank)

**Formula #2** - $\pi \times D \times H$ = 3.14159 x 5 x 10
= 157.08 square feet (total <u>area</u> of tank <u>wall</u>)

157.08 x 15.3 pounds = **2403.32 pounds** (<u>weight</u> of tank <u>wall</u>)

Conclusion:
**2403.32 + 600.83 = 3004.15 pounds (weight of entire tank)**

# 11
## MORE VALUABLE SCRAP METALS

## NON-FERROUS SCRAP

Below is a guide of the prices that dealers were paying for scrap at the end of 1982, in cents per pound.

|  | Atlanta | Boston | Cleveland | Detroit |
|---|---|---|---|---|
| No. 1 Copper | 46 | 46 | 44 | 45 |
| Yellow Brass | 25 | 27 | 27 | 28 |
| Aluminum | 21 | 19 | 19 | 19 |
| Lead | 8 | 11 | 10 | 9 |
| Lead Battery Plates | 6 | 6 | 7 | 6 |

|  | Los Angeles | New York | Philadelphia | San Francisco |
|---|---|---|---|---|
| No. 1 Copper | 46 | 47 | 47 | 50 |
| Yellow Brass | 31 | 27 | 27 | 29 |
| Aluminum | 24 | 19 | 20 | 21 |
| Lead | 11 | 11 | 11 | 10 |
| Lead Battery Plates | 5 | 6 | 6 | 5 |

Below, some averages of other metals (1982 prices, in cents per pound)

| Nickel | 115 | Monel (Nickel Alloy) | 70 |
|---|---|---|---|
| Zinc | 8 | Zinc Die Castings | 10 |
| Pewter | 250 | Auto Radiators | 33 |

We will approach each metal individually noting its characteristics, the likely uses, where it can be found, and methods of separation.

Methods of separation is important because for best prices the metal must be clean and free from other metals, attachments, bolts, screws, etc. This condition is imperative if you are selling to a smelter.

There are certain metals which would be costly to segregate. These mixed metals are sold to refiners, who know the exact melting point of each metal and bring their furnaces up to that temperature to extract that metal.

## ALUMINUM

Aluminum is a light, soft metal. It conducts electricity very well and is silvery white in color. Its symbol is Al.

Common uses are pots and pans, porches, car-ports, and house siding.

If mixed or alloyed with about 4% copper, it becomes tougher and is used as the main component in aircraft construction because of its lightness.

It is used for truck bodies, trailers and mobile homes for the same reason. As the price of copper continues to rise, aluminum is replacing it as a conductor of electricity in certain instances.

It can be cast in various shapes such as auto pistons and gearboxes. Wheels, aircraft and boat propellers are just a few of the thousands of uses for aluminum.

It is easily cut with an abrasive wheel or disc, melted with a torch or cut with a hacksaw, chisel or axe.

It has a low melting point so that items which are difficult to segregate, for example, aluminum which has steel bolts or screws attached, can be cleaned of these contaminations with the use of a torch.

Highly contaminated aluminum can be placed in a gas- or oil-fired furnace, allowing the aluminum to run out, leaving behind the steel.

Removing the insulation from aluminum cables is all that is needed to produce first class aluminum scrap.

## BRASS

Yellow brass is an alloy of 70% copper and 30% zinc. It is known as 70/30 brass and is sometimes used for cartridge cases and condenser tubes. It is obvious in the home when it is used to make decorative objects.

It is a little harder than aluminum, but it can be cut or worked easily.

## RED BRASS

Red brass is an alloy of about 83% copper and 17% zinc. It is used for bearings, valves, and various castings. Because of its higher copper content, it is worth more than yellow brass.

Red brass is almost always cast into various forms or shapes and after a while can be identified by its shape. Examples are pipe valves and bearing shells. A sharp blow with the edge of a file will show its true color.

## BRONZE

Bronze is usually 95% copper, 4% tin and only 1% zinc. It is used mainly for coinage. Bronze bells can contain up to 30% tin. The high prices of both copper and tin make this a valuable scrap metal. If lead is added it is useful for heavy duty bearings.

## MARINE BRONZE

As the name suggests it is used for some ship fittings. Phosphor up to 0.3% is added to deoxidize, where resistance to corrosion is required. For example, boiler fittings, gearwheels, propellers, shafts, rudders and other parts exposed to sea water.

High strength bronzes are made by adding small parts of manganese, iron, and aluminum to a 60/40 copper/zinc composition.

The dealer only offering small quantities of brasses would probably only separate into two grades, yellow and red.

Only when ton lots are available would it be worthwhile for further grading.

## COPPER

Copper is the basis for all brasses. Its symbol is Cu. It is soft with a high electrical and thermal conductivity and has a good resistance to corrosion, hence its use in water pipes and electrical cables. The color is red.

Copper pipes can be recovered from homes, mills, shops, manufacturing facilities, bottling plants, breweries, and chemical plants, in fact, everywhere where liquids are used.

It can be cut with a hacksaw, an abrasive disc, or it can be sheared. It needs too much heat for economical melting with a torch.

Almost all electrical cables with copper cores can be cleaned and made ready for sale by burning off the insulation.

Some wires or cables will burn (or at least the insulation will) without the need for other fuel. Some cables including those that are armored or protected with a steel outer covering will need to be placed on top of an existing fire. Heavy cables that are protected by steel should be cut into short lengths so that the fire can penetrate more easily.

Some cables have steel tape protection; this is twisted around the cable during manufacture. Others have a series of steel wires running lengthways but in a slightly spiral fashion. Some cables will also have a lead sheath; this will melt off in the fire and can be collected from the ashes later.

When preparing copper cable for sale, it is best to remove it from the fire while it is still hot so that it does not settle into the molten lead. We can also get a higher price for our copper wire if we plunge it into water when it is still hot. This will remove the ashes and dirt to give it a clean bright copper look.

Copper can be recovered for sale for instant cash from a variety of sources.

Here's a YouTube link of a guy explaining how much he gets for his pile of copper:

http://youtu.be/2StH2RoEFuU

And here's another YouTube video of the guy stripping copper out of radiators (Rated R for language!):

http://youtu.be/YkRc9LaoOkE

And on a similar scale, here's "Scrapper Girl" explaining her methods!

http://youtu.be/jcBDWyyExDQ

Electric motors and generators can supply good quantities of copper. The best from our point of view are **DC (direct current) motors**.

These have field coils which are fastened to the inside of the motor case, usually by bolts. If the ends of the motor are removed and the armature removed, the field coils can be extracted by giving each a sharp blow with a hammer, which will shear the holding bolts. The armature is the rotating part of an electric motor or generator. The steel cores can then be knocked out of the coils and the coils are now ready for burning with the armature.

The armature is also wound with copper, the ends of which are connected to the copper commutator. (The copper segments that are connected to the copper windings located at one end of the armature are known as the commutator.)

Before the armature is placed on the fire it is best practice to cut the windings close to where they leave the steel center section. If this is done with a hacksaw or an abrasive cutting disc, when the armature has cooled down you will find that the windings are easily detached.

The commutator segments are easily removed by unscrewing a locking nut on the end of the shaft.

The remaining iron and steel is sold as #2 steel scrap.

**AC (alternating current) motors** are not quite as easy to strip; they do not have field coils as such. A good way to identify DC motors is to look for the screws or bolt threads on the outside of the motor or generator - you can bet if they are not visible it is an AC motor.

Some AC motors have the outer shell or stator wound with copper, again if the windings are cut first before firing it makes for easy stripping. Some rotors are wound similar to the DC motor but have copper or bronze slip rings on the end of the shaft. Others have solid copper bars running through, while others, mainly induction motors, have aluminum in the rotors. This is usually melted out.

Generators and alternators are similar to motors in makeup but the process is reversed in use. Generators are driven or turned by other power to produce electricity, while motors are driven by electricity to produce power.

Some copper conductors of electricity are not insulated but are always placed out of reach to avoid the possibility of accidents. Nevertheless accidents do happen when a mobile crane driver, new to the area, touches them with his jib. Children climb pylons or poles carrying cables, or in a factory a long metal object touches the overhead traveling crane wires. Safety is key.

Some weights of copper per foot of cable are given further in this book. These will assist a cable buyer in knowing its instant cash value once it is burned or stripped.

I said stripped because not all insulated cables need to be burned. It is profitable to hand-strip certain cables by cutting the lead sheath with a broad chisel to get at the copper core. Some processors have machines where the cable is fed into adjustable rollers and a circular cutting knife splits the lead sheath - revealing the copper.

The lead recovered is better than common lead and if enough is available will command a premium price. The copper wire is also worth more, selling as #1 Bright wire unburned.

Going back to the overhead bare copper wire a moment - it should be snipped with wire cutters, coiled, and the ends secured with COPPER WIRE ready for the furnace. It is sometimes called HARD DRAWN wire and is probably two grades lower than #1 Bright but a little better than #1 burned.

Another type of bare copper conductor is the bus-bar or heavy strips or bars seen in electric generating plants, sub-stations, switchgear, circuit breakers, and other electrical equipment. A redundant generating plant or sub-station can produce thousands of dollars of instant cash; the copper can be recovered easily, using the methods described above.

The electricity is carried on power lines at a high voltage, or high pressure. The cables act like pipes - the greater the pressure, the thicker the pipe, or the heavier the copper. High voltage has to be reduced so that it can be used in the shops and homes. This is done with a transformer.

TRANSFORMERS are another good source of copper or instant cash. Transformers, as in electricity transformers, are usually made of steel cores wound with copper wire. They reduce the voltage and in doing so create heat. Smaller ones are air-cooled; larger ones are oil-cooled and are suspended in oil in steel tanks. The high voltage goes into the heavy copper coils and comes out at the voltage we can use.

The coils are lifted from the cases and are laid on their sides. They are wound around a laminated steel core. Most times, three coils and their laminated cores are joined by other laminations at the top and bottom. These can be removed by hand once the clamp that holds them together is unbolted. Laminations are light flat steel sheets bundled or stacked to form a heavy solid-looking core. The ones across the top and bottom are interspaced with the ones in the core but come away easily once the steel clamps are released. The separated cores can be pushed out of the coils and the coils made ready for firing.

After firing or burning off the insulation, water can be hosed over the hot coils to clean them further, for more instant CASH. Listed later are some weights of copper actually recovered from various sizes of transformers.

The actual generator at the power plant contains large quantities of copper. I have recovered up to 20,000 pounds from a single rotor. The turbine which drives the rotor can contain valuable metals in the blades. Also, the steam equipment used to drive the turbine will have condensers containing big weights of brass tubes and the heavy brass tube plates. There are cables from the generators to circuit breakers or switches and on to transformers.

The overhead cables, in rural areas, and underground in cities can contain up to 83,910 pounds of copper per mile. This is the figure for a cable made up of 217 strands of 0.152 inch diameter wire - 15,890 pounds per 1000 feet. A table for varying sizes appears later. Everything associated with electricity will produce valuable instant cash copper, even the telephone cables.

Other sources of copper are the sugar refining and sugar confectionary businesses. Copper kettles, boilers, and tanks are all available.

Breweries can have copper vats and fermenting tanks weighing thousands of pounds each as well as connecting non-ferrous pipes and valves.

Plumbers are replacing copper pipes every day, electricians and motor re-winders produce copper wire scrap. Heating and air conditioning contractors, mechanical engineers, the list is long. All can be persuaded to part with their scrap for ready cash.

Other sources are: roofing sheets, gutters, radiators, fire extinguishers, cooling and air conditioning units, electro-type shells and screens. Some old locomotives have heavy copper fire-boxes, some weighing up to three tons. Some boilers have copper tubes.

A steam turbine driving an alternating current generator

## TRANSFORMERS

The size or capacity of a transformer is specified in Kilo Volt Amperes (KVA).

If a winding is rated at 100 volts - 20 amps, the capacity is 100 x 20 = 2000 volt-amperes or 2 KVA.

A copper-wound steel core of a 50,000 KVA Transformer

The construction is similar on most 3-phase transformers, regardless of size.

The above picture shows the three copper coils which are wound on a laminated steel core. The coils and core all stand in a steel tank containing light oil which is a good insulation and cooling medium.

Most 3-phase transformers are made the same way, with the high and low voltage windings of the three phases on each leg. The yokes connecting the three coils are also made of laminated steel, each being approximately 0.014 inches thick.

When the steel channels which are bolted to each side, top and bottom, are removed, the yoke laminations begin to fall apart. We find that if the bottom channels are removed first and the whole assembly is lifted just above the ground, with a little persuasion the bottom yoke laminations fall to the ground. When they are cleared away, the assembly is lifted and then released or lowered quickly and the three coils usually drop to the ground.

They are then ready to be clipped into smaller sections and stacked ready for burning off the insulation. No other fuel is needed because of their previous immersion in oil. Just a little newspaper, a match and within an hour the fire has burned out and cooled sufficiently to move the clean copper either to a cash buyer or to a safe store place.

*(Editor's note: make sure the burning is done in a safe location!)*

Diesel-driven Direct Current generator

Notice the bolts holding the field coils to the outer shell, and the huge armature.

Following are actual copper weights recovered from the sizes of transformers listed. There is a variance between different makers.

| KVA Rating | | Weight of copper (pounds) | | Gross weight Inc. oil (pounds) |
|---|---|---|---|---|
| 5 | | 25 | | 210 |
| 10 | | 32 | | 290 |
| 15 | | 40 | | 300 |
| 20 | | 110 | | 705 |
| 25 | | 115 | | 750 |
| 50 | | 205 | | 1500 |
| 100 | | 300 | | 2100 |
| 150 | | 450 | | 2600 |
| 250 | | 560 | | 4100 |
| 250 | air cooled | 1200 | no oil | 4800 |
| 300 | | 600 | | 4600 |
| 400 | | 710 | | 5025 |
| 500 | | 850 | | 6100 |
| 750 | | 1250 | | 8960 |
| 800 | | 1300 | | 9310 |
| 1000 | | 1600 | | 11,200 |
| 2000 | | 3400 | | 23,600 |
| 7500 | | 6700 | | 53,000 |
| 12,500 | | 7650 | | 67,000 |

Below are weights of copper actually recovered from four-core cables.

| Diameter (inches) | Weight (Lbs/100 yards) |
|---|---|
| 0.03 | 140 |
| 0.06 | 288 |
| 0.10 | 484 |
| 0.20 | 944 |
| 0.30 | 1456 |

## Copper recovered from 23 DC electric motors:

Gross weight: 8 tons
Copper recovered: 1,920 pounds

## Weight of Copper Wires and Cables

| No. of Strands/ Dia. (in inches) of Wires | Pounds per 1000 feet | No. of Strands/ Dia. (in inches) of Wires | Pounds per 1000 feet |
|---|---|---|---|
| 217 / 0.152 | 15,890 | 19 / 0.105 | 653.30 |
| 169 / 0.133 | 9353 | 19 / 0.094 | 518.10 |
| 127 / 0.140 | 7794 | 19 / 0.084 | 410.90 |
| 127 / 0.125 | 6175 | 19 / 0.074 | 258.40 |
| 127 / 0.117 | 5403 | 19 / 0.066 | 230.46 |
| 91 / 0.128 | 4631 | 7 / 0.097 | 204.90 |
| 91 / 0.117 | 3859 | 7 / 0.087 | 162.50 |
| 61 / 0.128 | 3088 | 7 / 0.077 | 128.90 |
| 61 / 0.114 | 2470 | 7 / 0.069 | 102.20 |
| 61 / 0.111 | 2316 | 7 / 0.061 | 81.05 |
| 61 / 0.107 | 2161 | 7 / 0.054 | 64.28 |
| 61 / 0.099 | 1853 | 7 / 0.049 | 50.97 |
| 37 / 0.116 | 1544 | 7 / 0.043 | 40.42 |
| 37 / 0.104 | 1235 | 7 / 0.038 | 32.06 |
| 37 / 0.097 | 1081 | 7 / 0.030 | 20.16 |
| 37 / 0.090 | 926.30 | 7 / 0.024 | 12.68 |
| 37 / 0.082 | 771.90 | | |

## Dimensions and Weights

### Copper Wire
### Stranded Hard-Drawn Conductors

| Nominal Area (square inches) | Strands and Wire Diameter (inch) | Nominal Weight pounds per 1000 yards | kilograms per kilometer |
|---|---|---|---|
| 0.022 | 7 / 0.064 | 262.5 | 130.2 |
| 0.025 | 3 / 0.104 | 297.0 | 147.3 |
| 0.05 | 3 / 0.147 | 593.4 | 294.4 |
| 0.058 | 7 / 0.104 | 693.2 | 343.9 |
| 0.075 | 7 / 0.116 | 862.4 | 427.8 |
| 0.10 | 7 / 0.136 | 1186 | 588.1 |
| 0.15 | 7 / 0.166 | 1766 | 876.1 |
| 0.20 | 7 / 0.193 | 2387 | 1184 |
| 0.20 | 19 / 0.116 | 2351 | 1166 |
| 0.25 | 7 / 0.215 | 2963 | 1470 |
| 0.25 | 19 / 0.131 | 2998 | 1487 |

### Stranded Aluminum Conductors

| | | | |
|---|---|---|---|
| 0.0225 | 7 / 0.064 | 80.5 | 39.94 |
| 0.03 | 19 / 0.044 | 103.5 | 51.34 |
| 0.04 | 19 / 0.052 | 144.5 | 71.7 |
| 0.06 | 19 / 0.064 | 219.0 | 108.6 |
| 0.10 | 19 / 0.083 | 368.2 | 182.7 |
| 0.15 | 37 / 0.072 | 539.8 | 267.8 |
| 0.20 | 37 / 0.083 | 717.4 | 355.9 |
| 0.25 | 37 / 0.093 | 900.4 | 446.7 |
| 0.30 | 37 / 0.103 | 1104 | 548 |
| 0.40 | 61 / 0.093 | 1485 | 736.8 |
| 0.50 | 61 / 0.103 | 1821 | 903.7 |
| 0.60 | 91 / 0.093 | 2215 | 1098.7 |
| 0.75 | 91 / 0.103 | 2717 | 1347.8 |
| 1.00 | 127 / 0.103 | 3792 | 1881.6 |

## WIRE AND CABLE STRIPPING MACHINES

This machine is widely used on cables - from heavy wire-mesh-covered lead to light plastic-covered conductors with soft or firm body centers.

## MORE SOURCES OF CABLE CONTAINING COPPER

Shown above are heavy copper cables in the basement of a generating plant. Similar installations are found in heavy industrial plants and steel mills, etc.

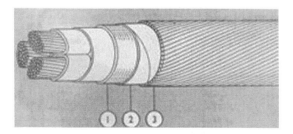

Above is a typical cable, showing 3 copper stranded cores each wrapped in paper. Then a covering of hessian, over that is (1) lead sheath (2) bitumen/hessian tape and (3) galvanized steel wire armor.

Single-core cable, lead-alloy sheathed.

## Copper Wire Weights
## (Single Wires)

| diameter (inch) | pounds per 1000 yards | diameter (inch) | pounds per 1000 yards |
|---|---|---|---|
| 0.500 | 2270.0 | 0.178 | 287.7 |
| 0.464 | 1955.0 | 0.176 | 281.3 |
| 0.460 | 1922.0 | 0.166 | 250.2 |
| 0.432 | 1695.0 | 0.162 | 238.3 |
| 0.400 | 1453.0 | 0.160 | 232.5 |
| 0.372 | 1257.0 | 0.1582 | 227.3 |
| 0.348 | 1100.0 | 0.152 | 209.8 |
| 0.324 | 953.3 | 0.147 | 196.24 |
| 0.300 | 817.3 | 0.144 | 188.34 |
| 0.276 | 691.8 | 0.137 | 170.45 |
| 0.252 | 576.7 | 0.136 | 167.96 |
| 0.232 | 488.6 | 0.131 | 155.83 |
| 0.2237 | 454.4 | 0.128 | 148.75 |
| 0.215 | 419.8 | 0.124 | 139.66 |
| 0.212 | 408.1 | 0.116 | 122.23 |
| 0.1938 | 341.1 | 0.1119 | 113.69 |
| 0.193 | 338.3 | 0.104 | 98.22 |
| 0.192 | 334.7 | 0.103 | 96.34 |

The photo above shows heavy copper cable being fed through a manhole. This cable is probably only paper and lead covered.

These cables are being laid in a duct which will be covered over later.

## Weight of Copper Sheet/Plate

| Thickness (inch) | Pounds per square foot | Thickness (inch) | Pounds per square foot |
|---|---|---|---|
| 0.300 | 13.950 | 0.064 | 2.976 |
| 0.276 | 12.834 | 0.056 | 2.604 |
| 0.252 | 11.718 | 0.048 | 2.232 |
| 0.232 | 10.788 | 0.040 | 1.860 |
| 0.212 | 9.858 | 0.036 | 1.674 |
| 0.192 | 8.928 | 0.032 | 1.488 |
| 0.176 | 8.184 | 0.028 | 1.302 |
| 0.160 | 7.440 | 0.024 | 1.166 |
| 0.144 | 6.696 | 0.022 | 1.023 |
| 0.128 | 5.952 | 0.020 | 0.930 |
| 0.104 | 4.836 | 0.0164 | 0.762 |
| 0.092 | 4.278 | 0.0148 | 0.688 |
| 0.080 | 3.720 | 0.0136 | 0.632 |
| 0.072 | 3.348 | 0.0124 | 0.576 |

## Weights and Melting Points

| Metal | Symbol | Atomic Weight | Melting Point (°C) |
|---|---|---|---|
| Bismuth | Bi | 209.00 | 271 |
| Lead | Pb | 207.21 | 327 |
| Gold | Au | 197.20 | 1063 |
| Platinum | Pt | 195.23 | 1774 |
| Tungsten | W | 183.92 | 3370 |
| Antimony | Sb | 121.76 | 630 |
| Tin | Sn | 118.70 | 232 |
| Cadmium | Cd | 112.41 | 321 |
| Silver | Ag | 107.88 | 960 |
| Arsenic | As | 74.91 | 850 |
| Zinc | Zn | 65.38 | 419 |
| Copper | Cu | 63.57 | 1083 |
| Cobalt | Co | 58.94 | 1495 |
| Nickel | Ni | 58.69 | 1455 |
| Iron | Fe | 55.84 | 1537 |
| Manganese | Mn | 54.93 | 1244 |
| Chromium | Cr | 52.01 | 1830 |
| Aluminum | Al | 26.97 | 660 |
| Magnesium | Mg | 24.32 | 651 |
| Beryllium | Be | 9.02 | 1284 |

## Weight of Copper Tubes

| Inside Diameter (inches) | Wall Thickness (in inches) | | | | | | |
|---|---|---|---|---|---|---|---|
| | 0.300 | 0.276 | 0.252 | 0.232 | 0.212 | 0.192 | Weight (in pounds per foot) |
| 1/2 | 2.90 | 2.59 | 2.29 | 2.05 | 1.83 | 1.71 | |
| 3/4 | 3.81 | 3.43 | 3.05 | 2.76 | 2.47 | 2.19 | |
| 1 | 4.27 | 4.26 | 3.82 | 3.46 | 3.11 | 2.77 | |
| 1 1/2 | 6.53 | 5.93 | 5.34 | 4.86 | 4.39 | 3.93 | |
| 2 | 8.35 | 7.60 | 6.86 | 6.26 | 5.67 | 5.09 | |
| 2 1/2 | 10.16 | 9.27 | 8.39 | 7.67 | 6.95 | 6.25 | |
| 3 | 11.98 | 10.94 | 9.91 | 9.07 | 8.24 | 7.41 | |
| 3 1/2 | 13.79 | 12.61 | 11.44 | 10.47 | 9.52 | 8.58 | |
| 4 | 15.61 | 14.28 | 12.96 | 11.88 | 10.80 | 9.74 | |
| 4 1/2 | 17.42 | 15.95 | 14.49 | 13.28 | 12.08 | 10.90 | |
| 6 | 22.86 | 20.95 | 19.06 | 17.49 | 15.93 | 14.38 | |

## Weight of Copper Tubes

| Inside Diameter (inches) | Wall Thickness (in inches) | | | | | | |
|---|---|---|---|---|---|---|---|
| | 0.176 | 0.160 | 0.144 | 0.128 | 0.166 | 0.104 | Weight (in pounds per foot) |
| 1/2 | 1.44 | 1.28 | 1.12 | 0.97 | 0.86 | 0.76 | |
| 3/4 | 1.97 | 1.76 | 1.56 | 1.36 | 1.21 | 1.07 | |
| 1 | 2.50 | 2.24 | 1.99 | 1.75 | 1.57 | 1.39 | |
| 1 1/2 | 3.57 | 3.21 | 2.86 | 2.52 | 2.27 | 2.02 | |
| 2 | 4.63 | 4.18 | 3.73 | 3.29 | 2.97 | 2.65 | |
| 2 1/2 | 5.70 | 5.15 | 4.61 | 4.07 | 3.67 | 3.28 | |
| 3 | 6.76 | 6.12 | 5.48 | 4.84 | 4.37 | 3.90 | |
| 3 1/2 | 7.83 | 7.08 | 6.35 | 5.62 | 5.07 | 4.53 | |
| 4 | 8.89 | 8.05 | 7.22 | 6.39 | 5.78 | 5.16 | |
| 4 1/2 | 9.96 | 9.02 | 8.09 | 7.17 | 6.48 | 5.79 | |
| 6 | 13.15 | 11.92 | 10.70 | 9.49 | 8.58 | 7.68 | |

## Weight of Copper Tubes

| Inside Diameter (inches) | Wall Thickness (in inches) | | | | |
|---|---|---|---|---|---|
| | 0.092 | 0.080 | 0.072 | 0.064 | Weight (in pounds per foot) |
| 1/2 | 0.66 | 0.56 | 0.50 | 0.44 | |
| 3/4 | 0.94 | 0.80 | 0.72 | 0.63 | |
| 1 | 1.21 | 1.04 | 0.93 | 0.82 | |
| 1 1/2 | 1.77 | 1.53 | 1.37 | 1.21 | |
| 2 | 2.33 | 2.01 | 1.80 | 1.60 | |
| 2 1/2 | 2.88 | 2.50 | 2.24 | 1.98 | |
| 3 | 3.44 | 2.98 | 2.68 | 2.37 | |
| 3 1/2 | 4.00 | 3.46 | 3.11 | 2.76 | |
| 4 | 4.55 | 3.95 | 3.55 | 3.15 | |
| 4 1/2 | 5.11 | 4.43 | 3.98 | 3.53 | |
| 6 | 6.78 | 5.88 | 5.29 | 4.69 | |

## Weights of Brass Bars

| Size (inches) (diameter or across flats) | Hexagon (Measure across flats) | Square (Measure across flats) | Round (Measure diameter) | |
|---|---|---|---|---|
| 0.25 | 0.202 | 0.232 | 0.182 | |
| 0.50 | 0.805 | 0.930 | 0.731 | Weight (in pounds per foot) |
| 0.625 | 1.260 | 1.450 | 1.140 | |
| 0.75 | 1.810 | 2.090 | 1.640 | |
| 0.875 | 2.470 | 2.850 | 2.240 | |
| 1.0 | 3.230 | 3.730 | 2.920 | |
| 1.5 | 7.250 | 8.370 | 6.580 | |
| 2.0 | 12.880 | 14.880 | 11.690 | |
| 2.5 | 20.130 | 23.250 | 18.270 | |
| 3.0 | 28.890 | 33.490 | 26.300 | |

For copper, multiply weight x 1.05
For aluminum, multiply x 0.67
For magnesium, multiply x 0.80
For zinc, multiply x 0.84

ALUMINUM PLATE - 0.5 inch thick; weight varies from 6.912 to 7.272 pounds per square foot because of specification for use in aircraft, boats, etc.

STAINLESS STEEL PLATE - 0.5 inch thick weighs 21.663 pounds per square foot.

BRASS PLATE - 0.5 inch thick weighs 21.90 pounds per square foot.

# 12
## COMMON METALS – LEAD AND ZINC

### LEAD

Lead is a soft bluish white metal and has a low melting point. Its symbol is Pb.

It is well known to most people because of its weight and common occurrence in water pipes. It protects some cables from moisture as previously mentioned and is useful in sheet form on roofs and gutters.

It is used in chemical plants because of its resistance to acids. Tanks and vessels are lined with it.

It is used in auto, truck and marine batteries, making them well sought after in the scrap trade: another source of instant CASH. It is a cycle of collector to dealer, dealer to merchant, and then merchant to consuming works.

Lead is used in great quantities in nuclear plants for its shielding properties. It is made harder by the addition of antimony.

Lead is the base metal for most solders, type metals, and some bearing metals known as Babbitt metal. These are worth more than lead because of the addition of TIN in varying quantities. Pewter tableware and soda-fountain boxes contain about 85% tin (cash value about $2.50 per pound, 1983 prices).

Solders are bought according to their tin content. Plumber's solder is not as valuable as tinman's solder. Plumber's solder can contain

30% tin, while tinman's solder contains 45%. Type metal from print shops contains about 6% tin. Old soda siphon tops are a form of pewter; to get $2.50 per pound (1983 prices) they must be free from rubber and broken glass.

## ZINC

Zinc is a hard bluish white metal. It symbol is Zn.

Because of its resistance to corrosion, its principle use is in galvanizing, which is a coating on steel to prevent rusting. It is used with copper to make brass and is used extensively in die-casting, especially in the auto industry for radiator grilles and door handles, etc.

Some processors install gas or oil fired furnaces to melt out zinc-based scrap such as pumps and carburetors, door handles, etc. They use the same furnace for aluminum content items such as gear-boxes, pistons with seized rings, and connecting rods. In fact, most metal with a low melting point such as lead and solder can be recovered from this type of furnace. They are always in the market for mixed scrap such as this because the end product is an ingot they can sell to be used for remanufacture.

## FURNACES

# *13*
# LESS COMMON METALS AND WHERE TO FIND THEM

## NICKEL

Nickel is a silvery white metal resembling iron. Its symbol is Ni. It is magnetic, and is used for electroplating and coinage.

It is used extensively in chemical and food-processing and handling plants. It is alloyed with copper (cupronickel), silver, brass, and light alloys. It is also used with chromium to make stainless steels.

The composition of stainless steels can vary according to the amount of protection from corrosion that is needed.

Typical compositions are:

1) 18% chromium, 8% nickel

2) 25% chromium, 12% nickel

These are highly resistant to corrosive attack by organic acids, weak mineral acids, and atmospheric oxidation.

Nickel is added to steel, up to 6%, to increase the strength and to enable hardening to be carried out in oil instead of water.

Nickel-Chromium Steel - 4% nickel and 2% chromium, give a steel tremendous strength. It is used in aircraft engine parts and armor plate.

When nickel is added to copper, up to 30%, it is used for coinage, condenser tubes, bullets, and turbine blades.

## MONEL METAL

MONEL METAL is a nickel base alloy containing 68% nickel, 29% copper, and 3% manganese, silicon, iron, and carbon. It has a high tensile strength and a high resistance to corrosion. Sources are: propellers, pumps, turbines, food handling and processing. See identification further on.

Editor's note: Other uses for Monel metal can be found here:

http://en.wikipedia.org/wiki/Monel_metal

## HIGH SPEED STEEL

Most stainless steels contain only 5% nickel but the chromium content varies from 10% to 20%. A good stainless steel is called 18-8 and contains 18% chromium and 8% nickel.

Tool steels are mentioned here because of their values as scrap.

HIGH SPEED STEEL is steel hardened by the addition of 12% to 18% TUNGSTEN, up to 5% chromium and about 0.5% carbon, plus very small amounts of VANADIUM and MOLYBDENUM.

It will keep its hardness at a low red heat making it useful for cutting tools in lathes, drills, etc. Various ratios of the above metals give us different uses for the tool steel.

Other ratios are: (tungsten/vanadium/molybdenum) 18/4/1, 6/5/2, and 9/2/1.

Some tools are only tipped with a hard cutting edge. The tips of TUNGSTEN CARBIDE are brazed in position. The tips are worth about five times more than high speed steel, about $3.50 to $4.00 per pound 1983 prices, (or about $5.50 per pound September 2012 price).

## 1980 Prices of These Types of Metals:

| Bismuth | 1.25 lb |
| Cadmium | 1.20 lb |
| Cobalt | 7.25 lb |
| Mercury | 1.50 lb |
| Molybdenum | 6.50 lb |
| Tungsten | 3.50 lb |

Here is a gasoline-powered cut-off saw with an abrasive disc. It is very useful on copper pipes and cable.

# 14
# IDENTIFICATION OF METALS

## COPPER

Color alone is usually sufficient but a magnet, if attracted, shows that the object is only copper plated. Best to chip or file to see red color.

## YELLOW BRASS

As the name implies, is yellow, can be chrome plated. Manganese Bronze is also yellow but is harder than brass and is mainly used for marine fittings, shafts, propellers, and rudders, etc. Again, best to chip or file for more certain identification.

## RED BRASS

Again, color speaks for itself, the red comes from more copper in the alloy so that the value is between copper and yellow brass.

## ALUMINUM

Soft, will scratch with a nail, non-magnetic and white in color. A solution of silver nitrate (0.5% in water) will remain clear if placed on a clean section.

## MAGNESIUM

It is similar to aluminum in appearance, and will give a dark gray to black sign in the above silver nitrate solution. Flakes, chippings, or filings will flare or burn brightly (hence its use in wartime incendiary bombs). It is used in engine blocks and wheels where lightness is important. It is lighter than aluminum.

## LEAD

Its heavy weight, gray color and softness are its identifying characteristics. Once handled it is never forgotten. It is non-magnetic.

## ZINC

Nitric acid applied to filings will produce effervescence and brown smoke.

## NICKEL

To test for nickel in alloys, a solution must be made up as follows:

1 gram of DIMETHYLGLYOXIME is dissolved in
50 ml of GLACIAL ACETIC ACID
Then:
10 ml of DISTILLED WATER is added, also
30 ml of CONCENTRATED AMMONIUM HYDROXIDE SOLUTION (NH4OH)

The mixture is thoroughly stirred until all the salts have dissolved. Then, 10 grams of AMMONIUM ACETATE are added.

This made-up solution is used in conjunction with acids when testing for nickel. It will produce a pinkish red color to a surface that has had acid applied.

## TIN

Nitric acid applied to filings will produce yellow smoke and a yellow color on the surface. Most of these tests are best carried out when the results can be compared with known samples, until you are familiar with the reactions and colors. **A successful scrap dealer always retains known samples for future comparisons.**

All surfaces to be tested must be cleaned with a file, emery paper, or grinding wheel. No grease, oil, or coating should be present and all plating must be removed.

A dealer will be able, after a short while, to tell one metal from another merely by recognizing its shape, color, and former use. All that remains is to present it to the buyer in clean worthwhile parcels. It is a commodity always in demand. There is INSTANT CASH waiting in every city for your collection!

## Identification Using a Grindstone (Spark Tests)

| Metal | Color of Spark | | Volume of Sparks |
|---|---|---|---|
| | near wheel | near end | |
| cast iron | red | straw | small |
| wrought iron | straw | white | large |
| steel | white | white | large |
| high speed steel | red | straw | small |
| nickel | orange | orange | very small |
| tungsten carbide | light orange | light orange | extremely small |
| copper | light orange | light orange | none |
| brass | light orange | light orange | none |
| aluminum | light orange | light orange | none |

# *15*
## MAGNETIC TESTING

This method is limited and is used merely to determine whether or not a metal belongs to the following group:

IRON - NICKEL - COBALT

All are attracted to a permanent magnet.

If the percentage of the above in an alloy is small then magnetic pull will not be perceived unless the magnet is hanging on a string.

# 16
## GOVERNMENT SCRAP SALES

A regular source of supply is the Defense Logistics Agency (DLA).

They are located at:

> DLA Disposition Services
> Hart-Dole-Inouye Federal Center
> 74 Washington Ave
> Battle Creek, MI 49037-3092
>
> Toll Free: 1-877-DLA-CALL (1-877-352-2255)
> Commercial: 1-269-961-7766

Additional sites are located around the world. A request to the above office will bring offers of varying amounts of scrap. You will be asked which parts of the US you are interested in. (Scrap is available in all parts.) A few typical offers are listed below.

COPPER-BEARING METALS, SCRAP: Consisting of electric motors, starters, alternators, generators, armatures and refrigeration compressors, with ferrous and non-ferrous attachments not to exceed 15% of total weight.
8500 POUNDS

COPPER, SCRAP: Consisting of tubing, pipes, strippings, fittings, cooling coils, rain gutters, wire wound relays and vehicle radiators, with ferrous attachments not to exceed 10%
2255 POUNDS

MISCELLANEOUS METALS, SCRAP: Including heaters, gauges, switches, electric fuel pumps, fans and rectifiers. Metallic content includes aluminum, copper, brass, stainless steel and steel. No-metallic content includes glass, plastic and rubber.
24,660 POUNDS

STEEL, HEAVY, PREPARED & UNPREPARED, SCRAP: Including axles, pipe, sheet steel, angles, gears, platforms, braces and baffles, with non-ferrous attachments and foreign material not to exceed 5% of total weight.

41 GROSS TONS

ALUMINUM, WRECKED AIRCRAFT AND IRONY SCRAP: Consisting of residue from A10 and A37 aircraft, including engine parts, with foreign and non-metallic materials not to exceed 15% of total weight.

6000 POUNDS

STAINLESS STEEL, MAGNETIC AND NON-MAGNETIC, SCRAP: Including coffee urns, toasters, panels, beverage boxes, steam tables, and vacuum jugs. Ferrous and non-ferrous attachments not to exceed 5% of total weight.15,000 POUNDS

COPPER, MIXED, UNSWEATED, VEHICULAR RADIATORS, SCRAP: With ferrous attachments, and foreign material not to exceed 25% of total weight.
26,700 POUNDS

ALUMINUM, IRONY, SCRAP: including obsolete and condemned helicopter blades.

16,000 POUNDS

AUTOMOTIVE SCRAP: Including heads, manifolds, wheels, crank shafts, gears, transfer case, pumps, alternators, camshafts, axles, springs, cylinders, starters, fans, rear ends, and fifth wheels.
17 GROSS TONS

MAGNESIUM, SCRAP: Including sheets, panels, covers and skins with ferrous and other non-ferrous attachments not to exceed 15% of total weight.
25,000 POUNDS

# 17
## PRECIOUS METALS

*PLATINUM   SILVER   GOLD*

## Sources of Platinum:

Catalytic converters
Contacts
Dental alloys
Dental scraps
Dental sweepings
Dental grindings
Jewelry scrap
Laboratory ware
Magneto points
Aircraft spark plugs
Thermocouple wire
Powders and pastes
Nozzles in nylon and acetate spinning

## Silver bearing materials are:

- Anodes
- Electric assemblies
- Silver/copper alloys
- Batteries
- Silver/cadmium alloys
- Silver/zinc alloys
- Silver/magnesium alloys
- Punchings
- Brazing alloys
- Brushes (electric motors)
- Bullion
- Chemical salts
- Electro-plated parts
- Contacts (points)
- Plated utensils
- Plated wire
- Powders
- Punchouts
- Clad bi-metal parts
- Coin silver
- Dental amalgam
- Contacts
- Films: litho- and X-ray
- Films: photographic
- Flake from hypo solution
- Plating hooks and nodules
- Jewelry
- Paints and paste
- Reproduction paper
- Resins
- Silver lined bearings in locomotives and aircraft
- Sludges
- Sterling
- Turnings

## Gold bearing materials are:

Placer gold (in rivers)
Plated parts
Plated wire
Circuit boards
Punchouts
Dental alloys
Dental scrap
Diodes
Plating filters
Flashings
Flakes
Resins - plating
Salts - chemical
Transistors
Jewelry
Filed scrap
Foil
Solutions
Plating hooks
Plating nodules
Sweepings
Contacts

# *18*
## SILVER COINS

U.S. coins dated 1964 and earlier, dollars, half dollars, quarters, and dimes have a large silver content. Based on a silver price (September 2012) of $34.50 per ounce, it is possible to realize $19 for every $1 face value.

Buyers are to be found in the Yellow Pages in every city. Kennedy half dollars 1965 – 1969 have a small silver content and are worth about three times face value.

## SILVER IN PHOTOGRAPHIC FILMS, LITHOGRAPHIC PLATES, AND X-RAY PLATES AND SOLUTIONS

Buyers of Silver Coins are listed as follows: (source: http://www.manta.com)

>Aria X-Ray Film Recycling
>Los Angeles CA 90210
>(214) 418-6720
>http://www.ariametals.com

>Liberty X-Ray Film Recycling
>Bedford TX 76022
>(888) 715-1022
>http://www.libertyxray.com

Rochester Silver Works
P.O. Box 15397, Rochester NY 14615
(585) 477-6434
http://www.rochestersilverworks.com

Karlan Service, Inc
189 E 7th Street, Paterson NJ 07524-1610
(973) 278-1015
http://www.karlanservice.com

Gulf States Silver, LLC
810 Fort Hill Dr., Vicksburg MS 39183
(601) 942-4242
http://www.gulfstatessilver.com

Commodity Resource & Environmental
116 E Prospect Avenue, Burbank CA 91502-2086
(800) 943-2811
http://www.creweb.com


X-ray film was bringing from $1.00 to $1.75 per pound of recovered silver. (September 2012 prices).


## HIGH VALUE AND EXOTIC METALS
## SEPT 2012 SPOT PRICES, PER OUNCE

|  |  |
|---|---|
| Palladium | $640.00 |
| Platinum | $1665.00 |
| Rhodium | $1100.00 |
| Iridium | $1050.00 |

Old mufflers sell for scrap according to the recoverable amount of Platinum in the catalytic converter - these can be obtained from muffler shops, and can be worth from $40 to over $200 (depending on the car or truck).

Check the market at eBay before selling. The catalyst itself is most often a precious metal. Platinum is the most active catalyst and is widely used. It is not suitable for all applications, however, because of unwanted additional reactions and/or cost. Palladium and Rhodium are two other precious metals used. Platinum and Rhodium are used as a reduction catalyst, while Platinum and Palladium are used as an oxidation catalyst.

# 19
# TESTING VARIOUS METALS

## SILVER

Place a drop of nitric acid on the sample. Allow it to react for about a minute. Place a drop of water on the acid spot, then adding a couple of crystals of common (table) salt to this liquid will produce a curded white look if the sample is silver.

## PLATINUM

This metal will not darken if heated and melts cleanly without oxides forming. The molten globule is smooth and white. Should it have a crystal pattern on its surface, it is probably iridium. A gas torch can be used for this test.

## GOLD

A sample is heated in a small beaker containing nitric acid. If it retains its color, it is gold. Whether it is only plated can be determined by cutting or filing. A quick test for a ring or chain is a drop of nitric acid for no reaction if gold is there. If the item is only plated, this test only shows that the outside is gold.

## SILVER PLATED OBJECTS

Nitric and hydrochloric acids will produce a milky white precipitate in the solution on a silver plated surface. Under the plating, if the base metal has a copper content, nitric acid will turn green.

If the base is aluminum it will be light in weight, and if the base metal is steel, it will be magnetic.

## NICKEL SILVER

Nitric acid will give a green color and a puff of brown smoke. It will turn pink when washed off with water.

# 20
## SCRAP DEALERS

Special thanks to http://www.manta.com, which has over 303,000 scrap metal listings on its website at the time of this writing, click the link to find dealers in your area.

---

Thank You for purchasing this book. Please feel free to leave a review over at Amazon.com and provide your feedback. It greatly helps other people in their decision to buy this book.

http://www.amazon.com/dp/B009TDDD2G

Typesetting, Design/Layout
and proofreading by
**Beesville Books**
www.beesvillebooks.com

Edited by Alan Gast

## OTHER BOOKS BY THIS PUBLISHER

*Easy Food Dehydrating and Safe Food Storage*

**Kindle version available at Amazon.com**
http://www.amazon.com/dp/B0093ZGX3Q

**Kindle version for our UK readers:**
http://www.amazon.co.uk/dp/B0093ZGX3Q

**Paperback:**
https://www.createspace.com/4009218

Visit Beesville Books for more information
at www.beesvillebooks.com

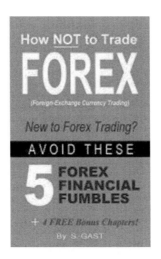

*How NOT to Trade Forex – Avoid These 5 Forex Financial Fumbles*

**Kindle version available at Amazon.com**
http://www.amazon.com/dp/B009SGIS1K

**Kindle version for our UK readers:**
http://www.amazon.co.uk/dp/B009SGIS1K

**Paperback:**
https://www.createspace.com/4055846

Visit Beesville Books for more information at www.beesvillebooks.com

~*~

How to Proofread Your Own Book:
A Five-Point Guide to Increase Your Credibility
(and Sales)

Kindle version available at Amazon.com
http://www.amazon.com/dp/ B00A6DSS7I

Kindle version for our UK readers:
http://www.amazon.co.uk/dp/ B00A6DSS7I

# ABOUT THE AUTHOR

Ken lives in a small, quiet town in Central Florida.
He lives with his wife Dode, and a calico cat
aptly named "Cali".

Made in the USA
San Bernardino, CA
05 December 2013